BROWN

by Melanie Mitchell

first step nonfiction

Lerner Publications · Minneapolis

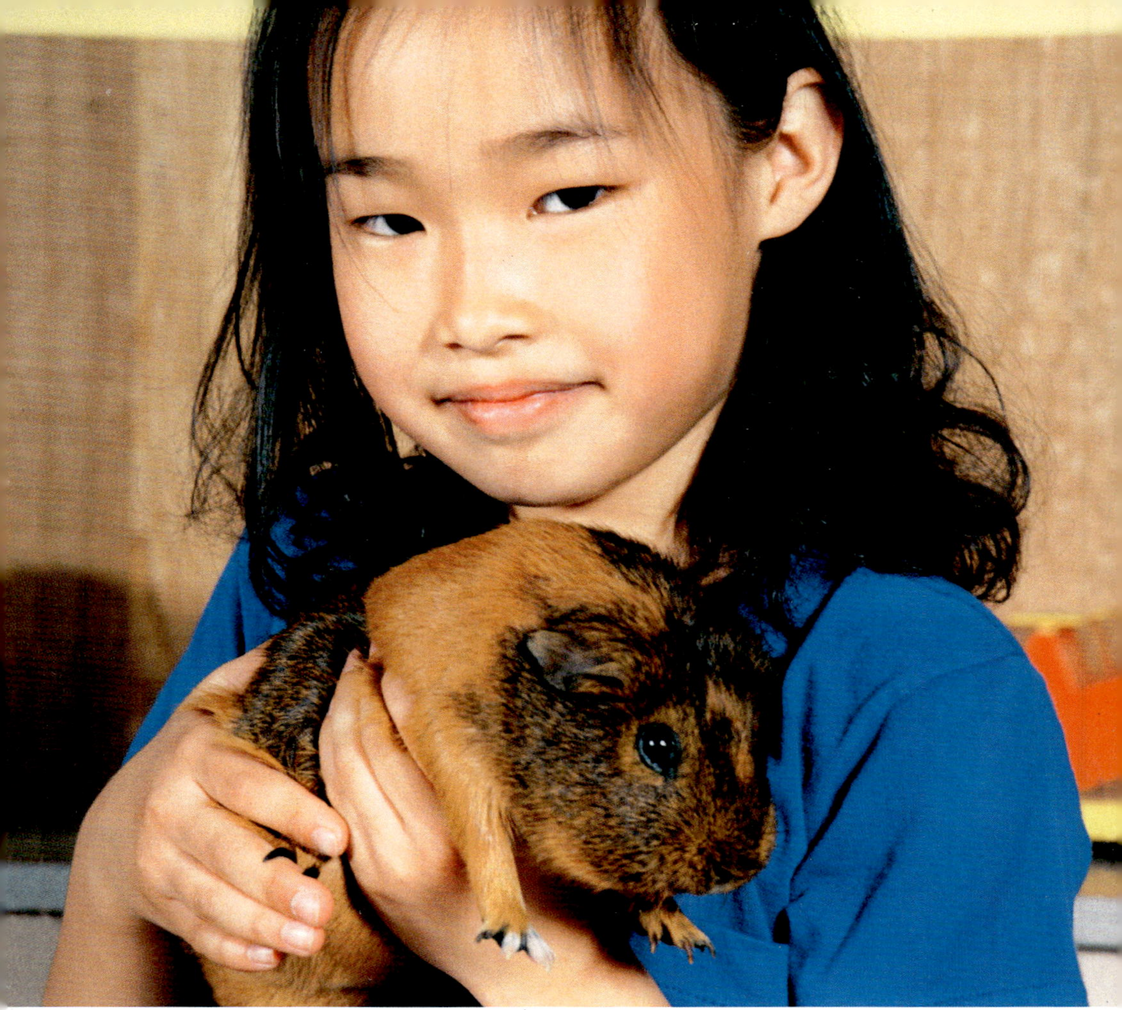

I see brown.

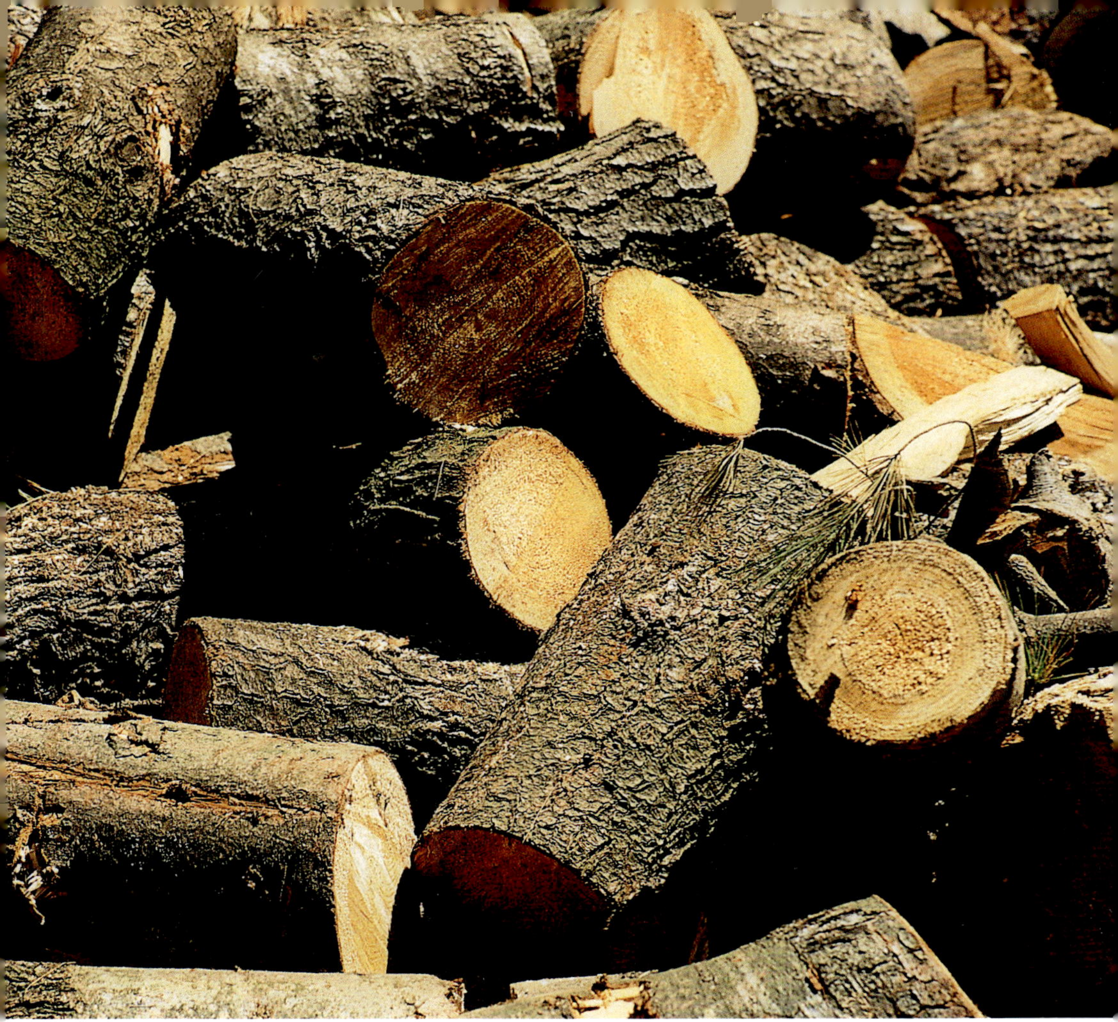

A log is brown.

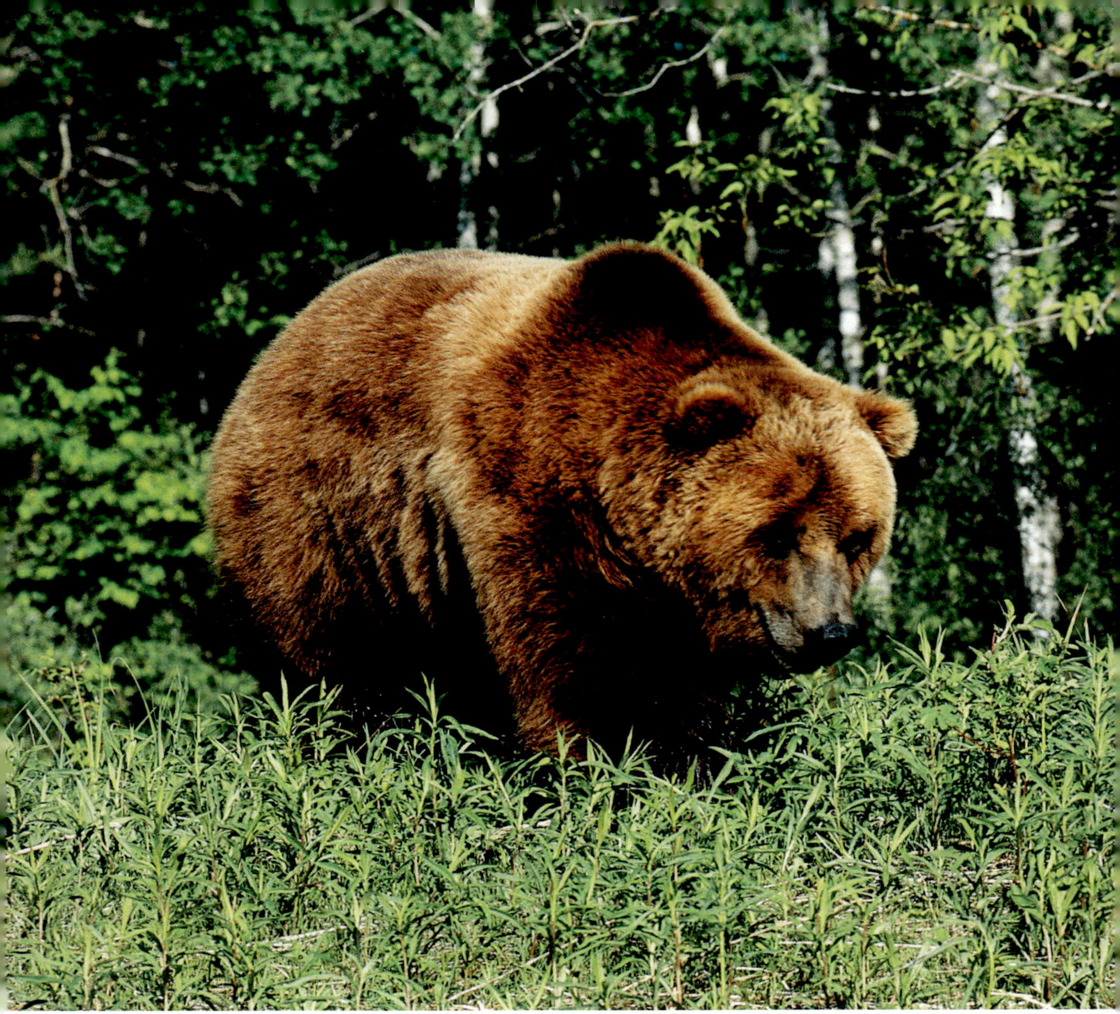

A bear is brown.

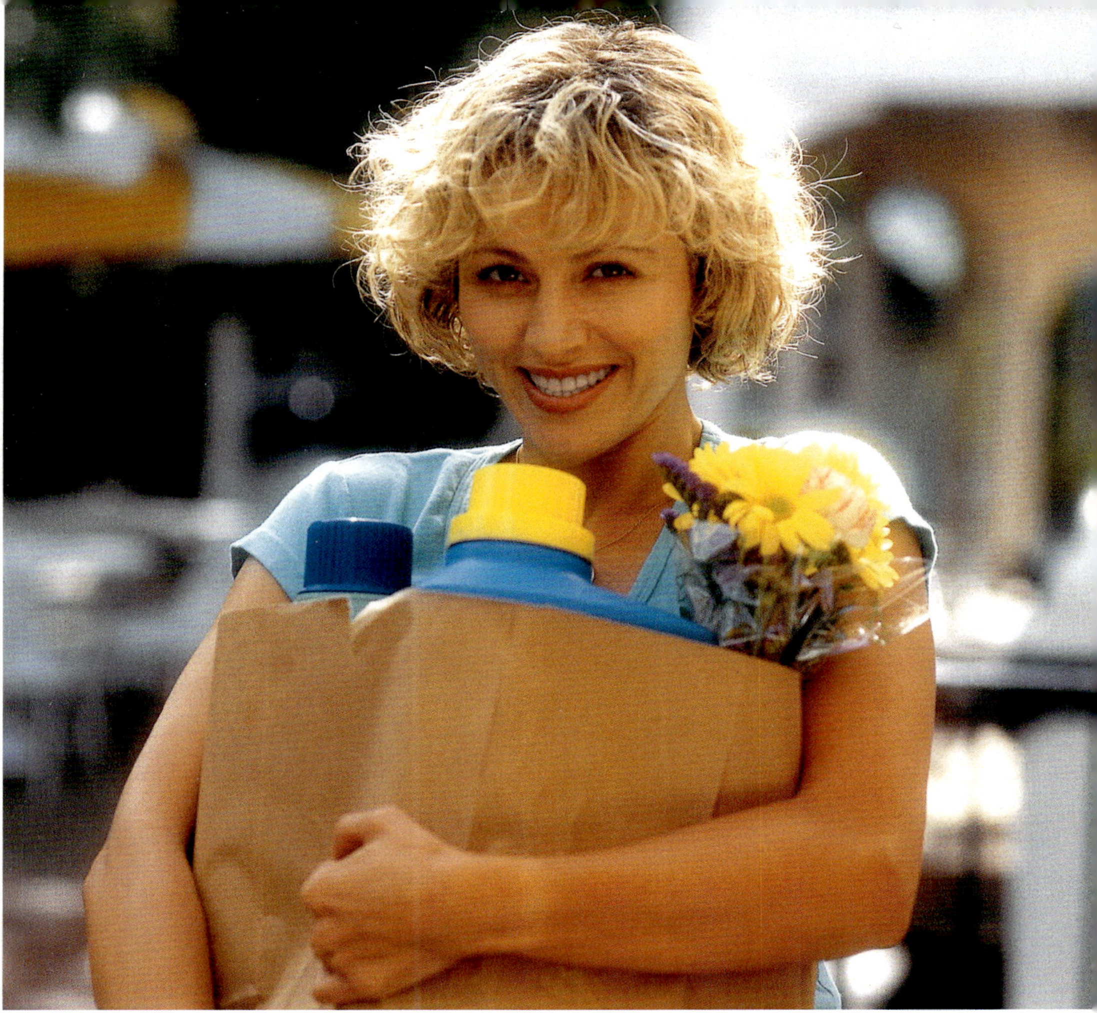

A bag is brown.

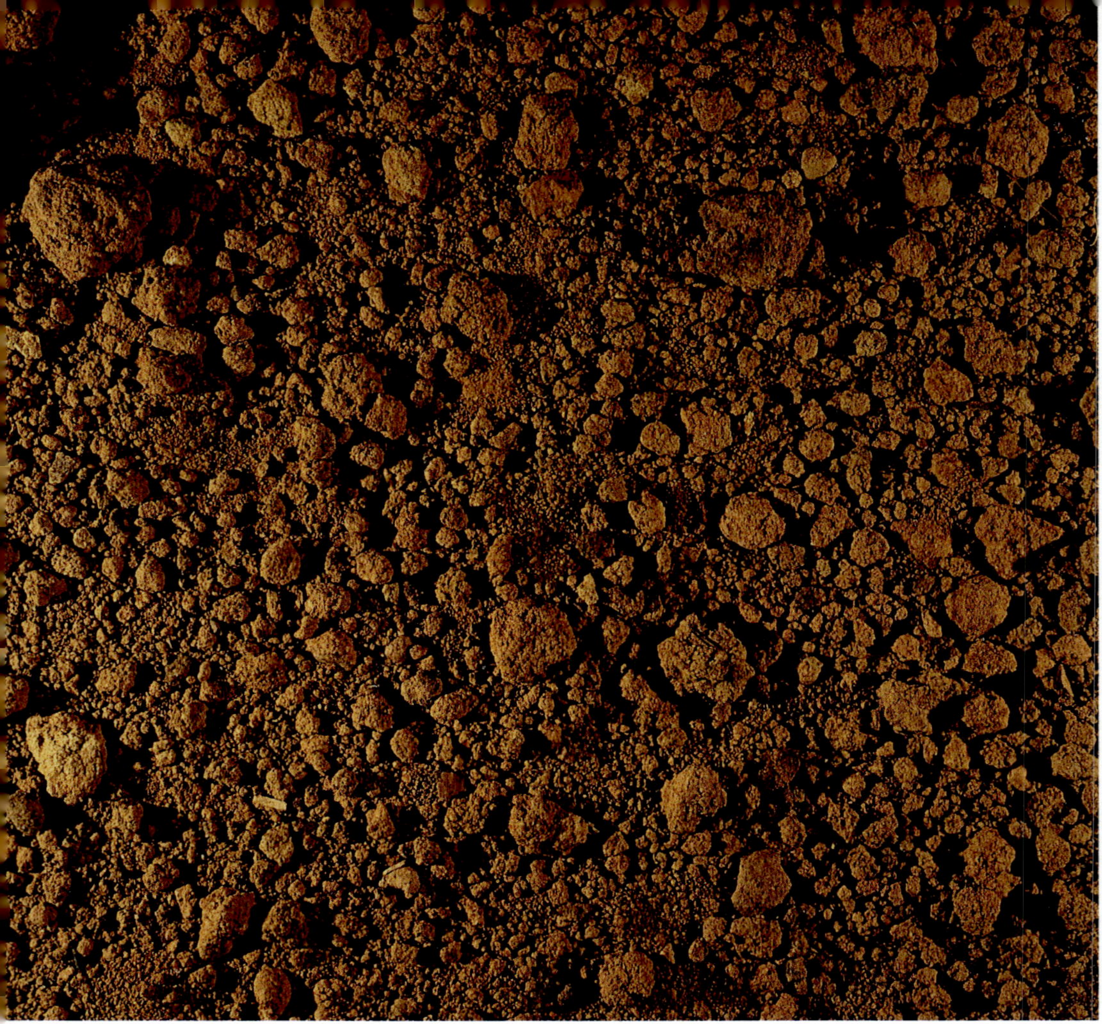

The dirt is brown.

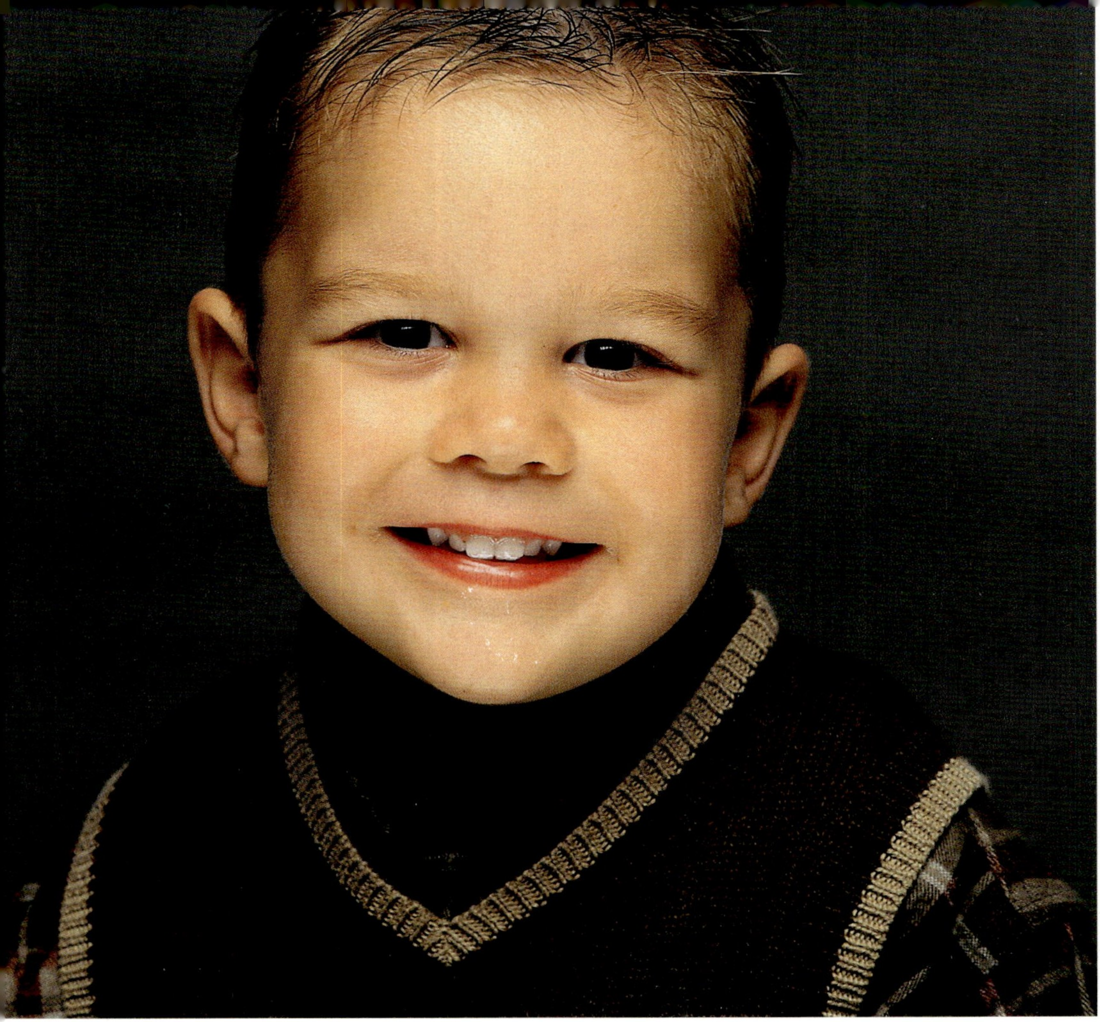

I like brown!

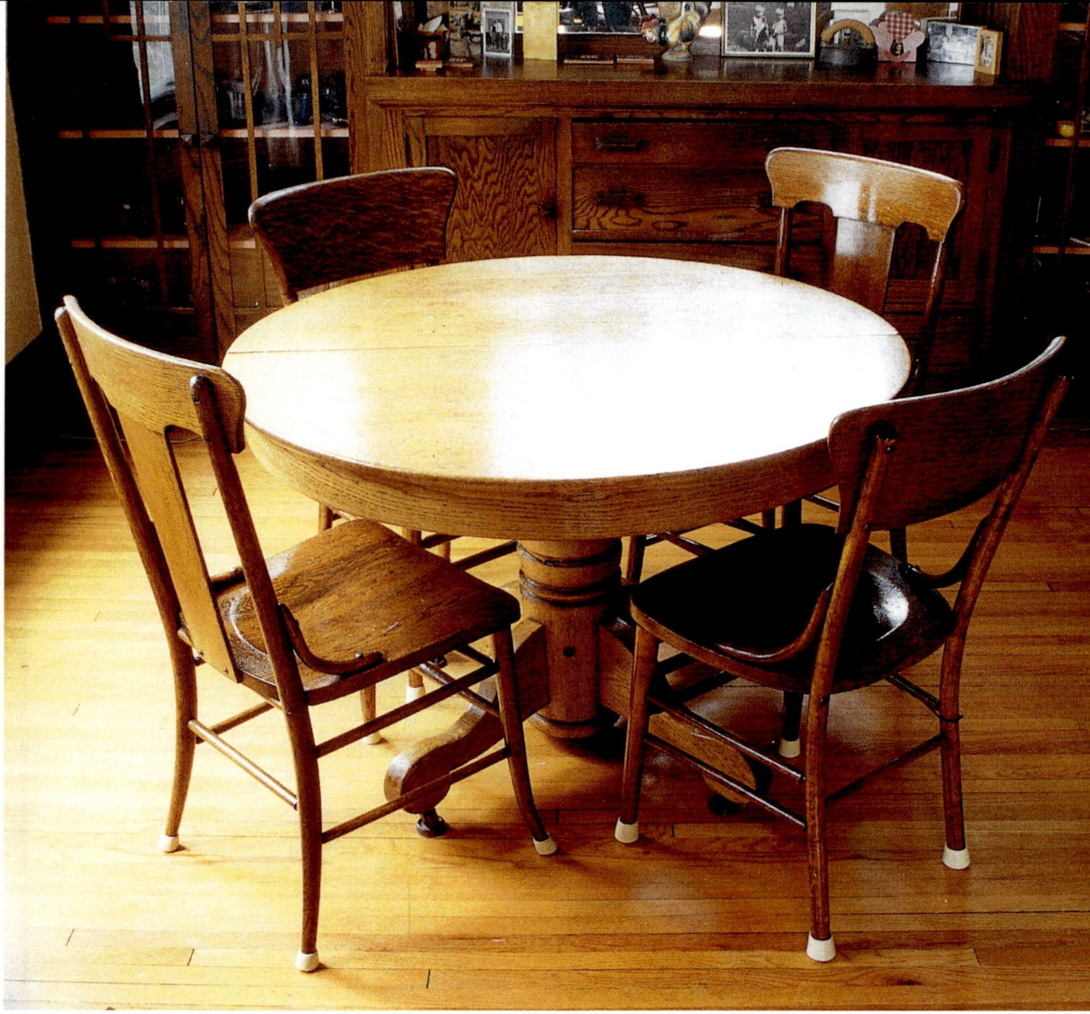

The table is brown.